Fernando Marini Kopp

ZBLAN:ER3+ energy up-conversion at low temperature

Fernando Marini Kopp

ZBLAN:ER3+ energy up-conversion at low temperature

Dissertation presented to obtain the title of
Master of Science - UEPG 2006

ScienciaScripts

Imprint
Any brand names and product names mentioned in this book are subject to trademark, brand or patent protection and are trademarks or registered trademarks of their respective holders. The use of brand names, product names, common names, trade names, product descriptions etc. even without a particular marking in this work is in no way to be construed to mean that such names may be regarded as unrestricted in respect of trademark and brand protection legislation and could thus be used by anyone.

Cover image: www.ingimage.com

This book is a translation from the original published under ISBN 978-613-9-61405-9.

Publisher:
Sciencia Scripts
is a trademark of
Dodo Books Indian Ocean Ltd. and OmniScriptum S.R.L publishing group

120 High Road, East Finchley, London, N2 9ED, United Kingdom
Str. Armeneasca 28/1, office 1, Chisinau MD-2012, Republic of Moldova, Europe
Printed at: see last page
ISBN: 978-620-7-62199-6

Table of contents:

I dedicate it to my wife Silvia and my children Helena and Pedro.
As a family, the logical framework of my existence and structure master of all my achievements.

Thanks

To God, absolute Lord of the sword and of time.

Copel - Companhia Paranaense de Distribuicao. for making this work possible in the person of Roberto Borges Pereira do Nascimento, Helder Cordeiro Barroso, Daniel Gueiber and Milton Hidekazu Iqueuti.

To my advisor, Prof. Dr. Gerson Kniphoff da Cruz, for the countless hours of dedication and help in guiding me through both my scientific initiation and this master's degree.

Dr. Cláudia Bonardi Kniphoff da Cruz for her help in revising the dissertation.

To MSc. Marcos Aurélio Viatroski for his help in revising the dissertation and providing samples for measurements.

Prof. Dr. Maria Cristina Terrile and Prof. Dr. René Aires de Carvalho, professors at the USP Physics Institute in Sdo Carlos, for making their laboratories and equipment available.

To the technicians of the resonance and cryogenics group at the USP Physics Institute in Sdo Carlos for their technical support in carrying out the measurements.

To everyone who directly or indirectly contributed to the completion of this journey.

Summary

The process of *upconversion* in fluoride glasses doped with rare earth elements (Er, Nd, Pr, Tm and Yb) has received a great deal of attention from researchers all over the world. The interest is mainly due to the applications of this system in optical amplifier devices and lasers with numerous technological applications. Fluoride glass (ZBLAN) is an amorphous material which, due to its advantages of low multiphonon relaxation, stability in atmospheric conditions and low optical attenuation in the visible region of the electromagnetic spectrum, is a good host material for rare earth ions. However, in this specific case, although it has been studied extensively, the ZBLAN:Er^{3+} system has some undefined optical properties. One of these is the effective participation of the glass matrix network in the up-conversion process. This paper presents *upconversion* and luminescence measurements at 2K with some definitions. Both spectra were obtained at the transition $S\,/^4{}_{32}$ -> $I^4{}_{i5/2}$ and serve to demonstrate that there is energy reabsorption in the *upconversion* process. The *upconversion* spectrum obtained from pumping the transition $I^4{}_{i5/2}$ -> $I^4{}_{9/2}$ with the aid of a Ti-Safira laser emitting at 800nm allows seven of the eight theoretically expected transitions to be verified. In order to verify the *upconversion* results obtained, luminescence measurements were carried out by pumping the $I^4{}_{i5/2}$ -> $F^4{}_{7/2}$ transition with the aid of an argon laser emitting at 488nm. From the luminescence, it is possible to completely identify the eight transitions expected in the $S^4{}_{3/2}$ -> $I^4{}_{i5/2}$ transition, in addition to confirming the positions of the emission lines obtained from the upconversion spectrum. We then propose a path for the excitation transitions leading to the *upconversion* process in the sample studied and an explanation (energy reabsorption process) for the disappearance of one of the emission lines in the *upconversion* spectrum.

Key words: *Upconversion*, ZBLAN, Erbium.

Chapter 1

1. Introduction

The optical phenomenon of up-conversion of infrared light energy into visible or ultraviolet in solids doped with rare earth elements has attracted a lot of interest in recent years due to the large number of applications already known. One of the most interesting applications is the production of short-wavelength lasers, driven by other commercially available infrared lasers [1]. The phenomenon of energy up-conversion is known in scientific circles as *"upconversion"*.

Upconversion is a process in which photons of a certain energy are absorbed followed by the emission of another photon with an energy greater than the energies of the individual photons absorbed by the system. Consider the 3-level system shown in figure 1.

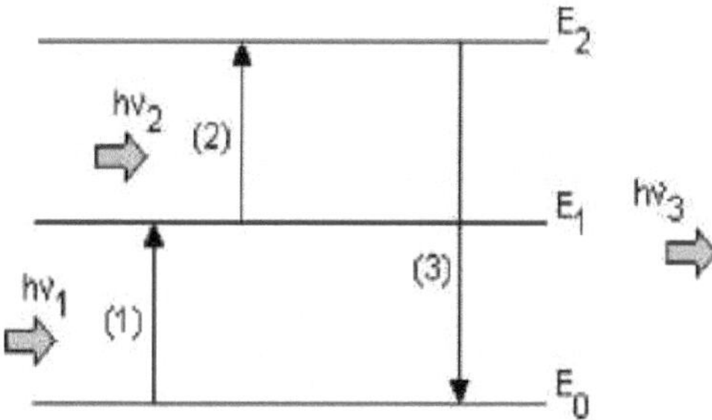

Figure 1 Simplified diagram of a 3-level energy system to visualize the *upconversion* process.

The system absorbs a photon of energy hv_1 from the excitation beam (source 1), passing from the energy state E_o to the energy state E_1 (1). The system then absorbs a new photon of energy hv_2 from the excitation beam (source 2), moving to the energy state E_2 (2). As the last stage of the process, the system decays from the energy state E_2 to the lower energy state E_o (3), emitting a photon with energy hv_3 (where $hv_1 < hv_3 > hv$)$._2$

In this work, a comparison will be made between the results of *upconversion* and luminescence, the latter of which will also need to be described below.

The phenomenon of luminescence (or photoluminescence) can be described as the absorption of a photon of a certain energy, followed by the emission of another photon of lower energy than the one absorbed. To visualize this process, figure 2 shows a diagram made up of 3 energy levels. The system absorbs a photon of energy $h\nu_1$ from the excitation beam, passing from the ground state with energy E_o to the excited state with energy E_2 (1). From the latter state, the system decays, by a non-radiative process, to the state of energy E_i (2). Next, the system decays again to the lowest energy state or fundamental state E_o (3), emitting a photon with energy h y (where h y<h v_i).

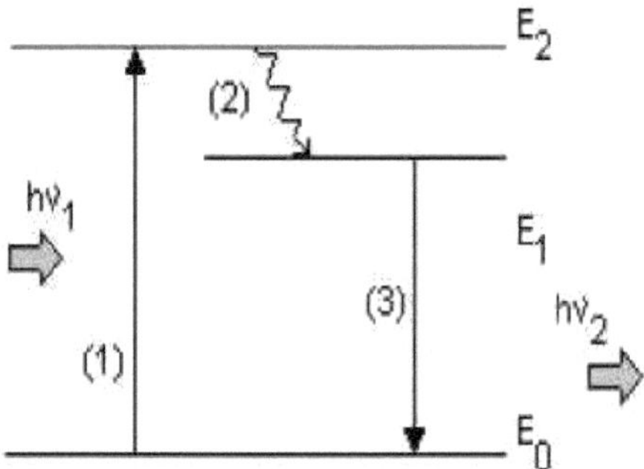

Figure 2 Simplified diagram of a 3-level energy system, to visualize the luminescence process.

Among the materials that allow the *upconversion* process to be achieved, glasses stand out as "host" materials for rare-earth elements. Various types of glass have been used over the last few years with a view to their practical use in optical and electro-optical systems. Examples of applications include various lasers [1, 2, 3, 4, 5, 6, 7, 8], color and/or 3D displays for the production of devices such as displays for various types of equipment [6, 7, 9, 10, 11, 12], sensors for auxiliary action in various control devices [1, 4, 6, 7,12], optical data storage such as optical CD or DVD players/writers [6, 7, 10, 11, 12], diagnostics and other medical applications such as surgery [10, 11, 12, 13], infrared transmitters and/or optical amplifiers [3, 8, 14] and signal amplifiers in telecommunications [15].

The various types of glass analyzed and researched by the scientific community include: fluorozirconates (ZBLAN) [2, 3, 5, 6, 8, 9, 11, 13, 17, 18, 19, 20], fluorinated [1, 14, 19, 21], oxides [4, 12], silicates [20], oxyfluorinated [22], germanium-based [20], phosphates [15] and chlorofluorozirconates [3].

While silica-based glass matrices have excellent optical transmission properties for wavelengths shorter than 2 um, other matrices are required for transmission of longer wavelengths in the mid- and far-infrared regions. Figure 3 reproduces the transmission spectrum for some of these different classes of materials [16]. It shows a comparison between 5 types of material, with the wavelength range in which they are applied associated with the percentage of transmittance of the material.

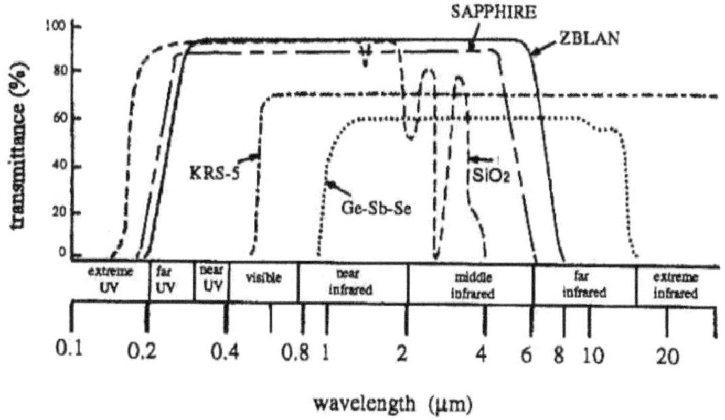

Figure 3: Transmission spectrum for some materials [16].

It can be seen that the KRS-5 material has optical applications from approximately 0.6 um to wavelengths greater than 20 um with a transmittance of approximately 70%. Ge-Sb-Se-based glasses have optical applications in the range of approximately 1 to 15 um with an approximate transmittance of 60%. Silicate glasses (based on SiO_2), which are the most common and cheapest, have optical applications from approximately 0.18 um to 1.5 um, continuously, with a transmittance of approximately 95% (from 1.5 to 4.0 um there are optical restrictions for some wavelengths). Sapphire has an optical application from wavelengths of approximately 0.2 um to approximately 6 um, with a transmittance of approximately 90%. And finally, ZBLAN, with optical application from wavelengths of 0.2 um to approximately 8 um, with an approximate transmittance of 95%, which is an advantage over the others. Other advantages for ZBLAN glass are the ease of manufacture directly linked to the temperatures involved [7, 17] and the lower processing costs linked to the energy required to produce it [17].

Despite the large amount of research and notable technological advances around the world in the use of optical fibers produced from various vitreous matrices, solid state spectroscopy presents challenges in the knowledge and application of the optical properties of the rare earth ions present in these vitreous matrices. One example of the need for progress in this area is obtaining greater efficiency in the phenomenon of *upconversion* in fluorinated glass matrix doped with rare earth ions.

The aim of this dissertation is to compare the emission spectra by up-conversion and luminescence, to identify the energy levels involved and to propose a possible pathway leading to up-conversion at $T = 2K$ for the transition $^{4}S_{3/2} \rightarrow {}^{4}I_{15/2}$. Thus The dissertation is divided into five chapters.

Chapter 2 discusses the sample, with the justification for using the ZBLAN glass matrix doped with the erbium element in its trivalent form Er^{3+}. Chapter 3 describes the systems used to obtain the experimental results. The results and their discussions are in Chapter 4. Finally, chapter 5 is reserved for the conclusions.

Chapter 2

2. *The vitreous matrix and the Er^{3+}*

2.1. *Justification for using the vitreous matrix*

Transparency, hardness at room temperature, mechanical strength, corrosion resistance and easy processing are all examples that have made glass one of the best known and most used materials in everyday life.

Glass can be defined in two ways. The first is associated with the conventional method of its preparation, i.e:

"Glass is obtained from the rapid cooling of a material in the liquid state, without the occurrence of crystallization, until it forms a rigid solid, by increasing its viscosity" [17].

The second is associated with atomic structure:
"A glass is a non-crystalline solid with no long-range order or periodicity in the atomic arrangement and which exhibits the glass transition phenomenon" [17].

The glass transition differentiates glasses from crystalline materials and can be understood with the help of figure 4, which shows the solidification diagram of crystalline materials and glasses. It shows that from a point A at which a material is in a liquid state and is cooled, its specific volume decreases uniformly within the defined region up to the crystalline material's melting temperature (*Tf*).

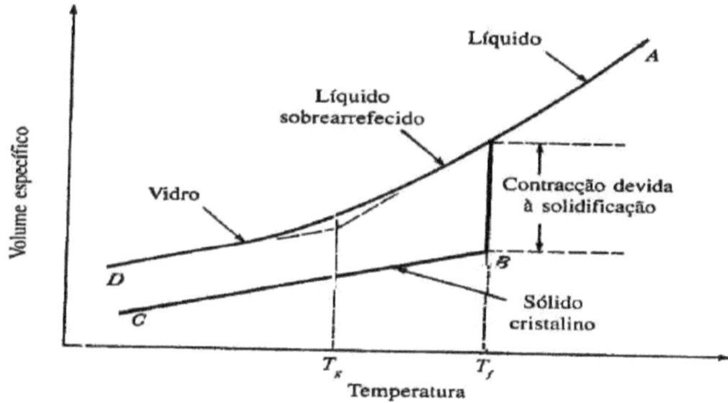

Figure 4 - Diagram of the solidification of crystalline materials and glasses showing the changes in specific volume. T$_g$

is the glass transition temperature for a given cooling rate. T_f is the melting temperature of the crystalline material [17].

If the cooling rate represented in figure 4 by the line between point A and temperature Tf is sufficiently slow, crystallization occurs at temperature Tf. The specific volume decreases abruptly up to point B and from this point the crystalline solid formed contracts uniformly with the drop in temperature, along the line BC.

If the aforementioned cooling rate is otherwise fast enough, crystallization does not occur in the Tf position. The cooled liquid becomes increasingly viscous, moving from a pasty state to a rigid solid state. In this narrow temperature range, the slope of the specific volume curve as a function of temperature decreases appreciably. The point where the two slopes of this curve intersect defines a turning point which is called the glass transition temperature T_g . Therefore, the glass transition temperature does not have a constant value for a given substance, but varies with the rate of cooling.

There is a temperature at which glass can crystallize. This temperature is identified as T_x and the greater the difference between this value and T_g the easier the glass is to process and therefore the more stable it is. ZBLAN fluoride glass has this stability characteristic [17]. The values involved are shown in table 1.

Table 1. Values of the temperatures of glass transition (Tg), crystallization (Tx) and fusion (Tf) obtained for the glass sample ZBLAN:Er^{3+} , using the differential calorimetry technique (DSC). The table also shows the difference between Tx and Tg.

Sample	T_g	T_x	Tf	$T - T_{xg}$
ZBLAN:1%Er^{3+}	264 C^0	335 0C	443 0C	71 0C

Source: Bibliographic reference [17].

Other characteristics that point to the use of the ZBLAN matrix are also noteworthy: accepts large quantities of dopants or additives [1, 19, 21], provides uniform distribution of rare earth ions within its structure [8, 17], low phonon energy [2, 3, 5, 11, 13,

14, 17, 18, 19], transparency for wavelengths in the mid-infrared to near ultraviolet region [13, 19, 21] and good mechanical medium for the *upconversion* phenomenon [6, 9].

2.2.0 Er ion^{3+}

Among the rare-earth elements, the one with the highest efficiency in the *upconversion* emission characteristic and therefore widely studied is erbium in the trivalent state (Er^{3+}) [11]. It is cited as being "rich in luminescence phenomena" [22], which certainly justifies its position among the most researched elements in this area. We can highlight the fact that the Er ion^{3+} has important characteristics: 1) It has energy levels of the same magnitude as other rare earths; 2) It is a good energy acceptor in the *upconversion* process; 3) It has an ionic radius similar to that of other rare earths, thus making it possible for the ions to be evenly distributed in crystal matrices [8].

2.2.1 The Hamiltonian of the Er ion^{3+}

The element erbium belongs to the lanthanide series characterized by the progressive filling of the *4f layer*. This innermost layer is "shielded" by the *5s^2* and *5p^6* layers. Thus, the *4f* electrons of erbium are weakly disturbed by their neighbors when the element is introduced into a solid. In the triple ionized state, the electron configuration is [Xe] *4f^{11}* .

The free ion Hamiltonian [24] can be written in the form:

$$H = H_O + H_{ee} + H_{SO} \qquad (1)$$

The H_0 component takes into account the kinetic and potential energy of the electrons. The *H component$_{ee}$* introduces the electrostatic interaction between the *4f* electrons. From this contribution we have terms that are characterized by the quantum numbers L and S. The last interaction to be considered is spin-orbit (H_{SO}), which together with the contribution of the electrostatic interaction are responsible for the energy level structure of the *4f* electrons of the free ion determined in table 2 [23].

Table 2. - Energy values (cm^{-1}) for the energy levels of the free Er ion^{3+} and in various crystalline matrices.

Term	Free ion	LaF$_3$	ErES^b	ErCl$_3$·6H$_2$O	LaCl$_3$	LaBr$_3$	Y$_2$O$_3$
$^4I_{15/2}$	0.0	0.0	0.0	0.0	0.0	0.0	0.0
$^4I_{13/2}$	6485	6480			6481	6475	6458
$^4I_{11/2}$	10,123	10,123	10,113	10,109	10,111		10,073
$^4I_{9/2}$	12,345	12,350	12,366	12,349	12,351	12,338	12,287
$^4F_{9/2}$	15,182	15,235	15,207	15,182	15,175	15,149	15,071
$^4S_{3/2}$	18,299	18,353	18,327	18,284	18,290	18,260	18,072
$^2H_{11/2}$	19,010		19,087	19,055			18,931
$^4F_{7/2}$	20,494	20,492	20,457	20,426			20,267
$^4F_{5/2}$	22,181	20,161	22,121	22,078	22,067	22,021	21,894
$^4F_{3/2}$	22,453	22,494	22,461	22,436	22,409	24,369	22,207
$^2H_{9/2}$	24,475	24,526	24,515	24,464	24,433		24,304
$^4G_{11/2}$	26,376	26,368	26,348	26,297	26,271	26,180	26,074
$^2G_{9/2}$	27,319	27,412	27,360	27,285		27,159	
$^2K_{15/2}$	27,584		27,660	27,649			
$^2G_{7/2}$	27,825	28,081	27,970	27,940			
$^2P_{3/2}$	31,414	31,501	31,480		31,384	31,284	31,186
$^2P_{1/2}$				32,630			
$^2K_{13/2}$			32,960				
$^4G_{5/2}$			33,250				
$^4G_{-/2}$	33,849	33,995	33,930				33,697
$^2D_{5/2}$			34,810				
$^4G_{9/2}$			36,370				

^{b}Er(C H$_{25}$ SO)$_{43}$ '9H O$_{·2}$
Source: Bibliographic reference [23].

The same table shows the central value of these levels in various materials. These energy values take into account an additional contribution to be included in the Hamiltonian of equation 1, the electrostatic interaction that arises between the ion and its neighbors when this ion is placed in a crystal lattice $(H)_{.cc}$

One fact to note in this table 2 is that the central value shown for the energy levels varies by around 100 cm^{-1}. In some cases this difference can reach 200 cm^{-1}, i.e. for the study of the Er ion^{3+} in a material we can consider the free ion as the starting point for identifying the transitions observed experimentally.

Figure 5 schematically represents the energy levels, showing the influence of each part of the Hamiltonian in equation 1 [24]. The order of magnitude of the separation between levels can be seen as the various parts of the Hamiltonian are considered. The figure includes the action of the crystal field $(H)_{cc}$.

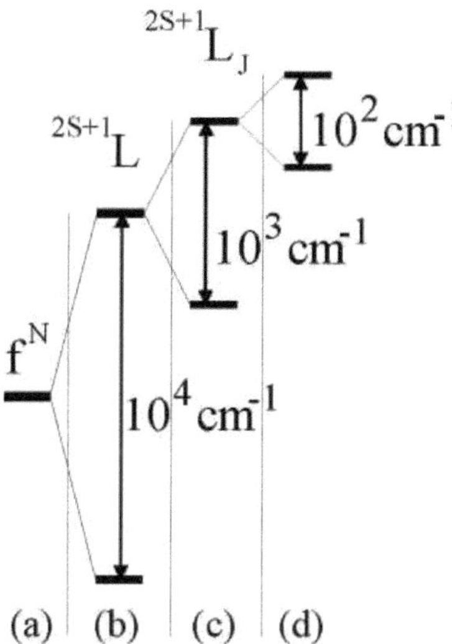

Figure 5 Energy level diagram illustrating the effect of the various terms in the Hamiltonian of equation 1:
(a) H_0, (b) $H_0 + Hcc$, (c) $H_0 + Hee + Hso$ and (d) $H_0 + Hee + H_{so} + Hcc$ with the order of magnitude values of the separations between levels indicated [24].

The level diagram shown in figure 6 was constructed using the data in table 2. In the diagram, the dotted arrows indicate the excitation transitions used. The emission transition studied is marked on the solid line.

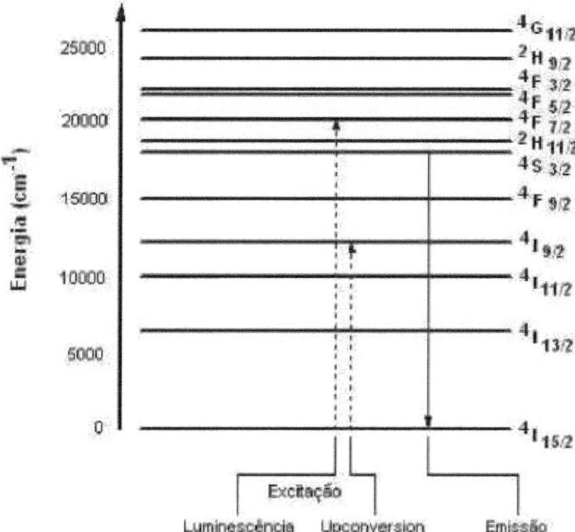

In principle, these transitions would be forbidden by the electric dipole because they are transitions that connect states of the same parity (Laporte's rule). However, due to the influence of the crystal lattice, as discussed in Judd and Ofelt's theories, they are allowed. The odd terms of the crystal field Hamiltonian and the coupling with the lattice or the interaction with states of different parity configurations, such as the excited configurations $4f^{(n-1)}5d$, are sufficient to produce the mixture of $4f$ states with different parities and allow these transitions [25, 26].

In the case of interest, we will consider that the existence of the crystal field around the Er ion^{3+} leads to the degeneracy of the multiplets in the maximum number of Stark levels. Under this condition, the expected breakdown of energy levels $^4S_{3/2}$ e $^4I_{15/2}$ is obtained from the table [27] below:

Table 3. Breakdown of energy levels for semi-whole J.

Type of symmetry	J →	1/2	3/2	5/2	7/2	9/2	11/2	13/2	15/2	17/2
Cubic		1	1	2	3	3	4	5	5	6
Other Groups		1	2	3	4	5	6	7	8	9

Source: Bibliographic reference [27].

Chapter 3

3. *Experimental procedures*

The aim of this chapter is to describe the experimental measurement systems. At this point, we would like to thank the Magneto-Optics Laboratory of the Physics Institute of Sao Carlos - USP, which allowed access and provided financial support for the purchase of the material used to carry out the measurements. Thanks also go to Prof. Dr. Younnes Messadeq, from the Chemistry Institute of Unesp in Araraquara, who allowed the use of his laboratory so that MSc. Marcos Viatroski could produce the vitreous sample studied.

To produce the sample, five grams of fluorides were melted in their respective molar compositions: 50% ZrF_4 + 20% BaF_2 + 4% LaF_3 + 5% AlF_3 + 20% NaF + 1%ErF_3 . The crude fluorides were melted in a platinum crucible at a temperature of 900 °C for 20 minutes in an air atmosphere. The molten product was poured into a tin mold at a temperature of 220 °C. The glass obtained was annealed for one hour and cooled slowly (approximately 10 hours) in order to reduce residual stresses [17].

Below is a list of the main equipment and accessories used in measuring systems, which will be described below:

Monochromator: Jobin Yvon - SPEX 1 m (net: 1200 lines/mm);

Laser: Lexel Laser INC. 516 Ar serial 12607;

Modulator: SR 540 Chopper - Stanford Research Systems Inc;

Lock-in amplifier: SR 530 - Stanford Research Systems Inc;

Cryostat: Intermagnetic immersion cryostat;

Oscilloscope: Tektronix 2221A;

Color filters: Oriel Corporation;

Flat mirrors and BK7 lenses;

Photomultiplier: Hamamatsu - Type R 636 - 10.

3.1. *Description of measurement systems*
3.1.1.   *Upconversion of energy*

Figure 7 shows the configuration of the experimental system used to obtain the *upconversion* spectra.

The first laser (*FA*) is an Argon laser, used to pump the Titanium-Sapphire laser (*FB*), which in turn is tuned to a wavelength of 800 nm. This laser beam is incident on the flat mirrors *E1* and *E2,* which direct the beam onto the sample. The converging lens *L1* concentrates the excitation light on sample *A*.

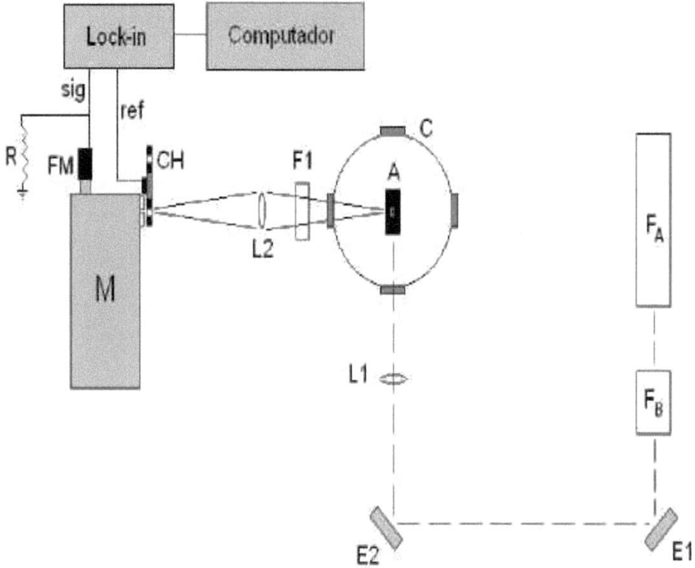

Figure 7 Representative diagram of the system used for *upconversion* measurements.

The light emitted by the *upconversion* is collected by lens *L2,* which focuses this light on the entrance slit of monochromator *M*. Before falling on the monochromator, this light beam passes through chopper *CH* and filter *F1* (a 690 nm low-pass filter that blocks the excitation light that may have been collected by lens *L2* and is directed to the detection system). The *FM* photomultiplier valve receives the light signal and converts it into an electric current that circulates in resistor *R*. The potential difference measured at the ends of this resistor *R* is proportional to the intensity of the light incident on the photomultiplier. The lock-in amplifier measures this potential difference. The *CH* chopper supplies the signal reference to the Lock-in amplifier, which is connected to the computer that stores the results.

3.1.2.   *Luminescence*

Figure 7 shows the configuration of the experimental system used to obtain the luminescence spectra.

The light source in this case is an argon laser (*F*), whose beam passes through the spatial filter (*FE*), which consists of a prism (*FP*), the *El* mirror and the slit. The prism (*FP*) separates the lines emitted by the *F* source, which works in multi-mode with wavelengths of 476.5 nm, 488.0 nm, 496.5 nm, 501.7 nm and 514.5 nm. All of these lines are incident on the flat mirror *El* used only to direct them towards the slit, which in turn is positioned so that only light with a wavelength of 488 nm can pass through.

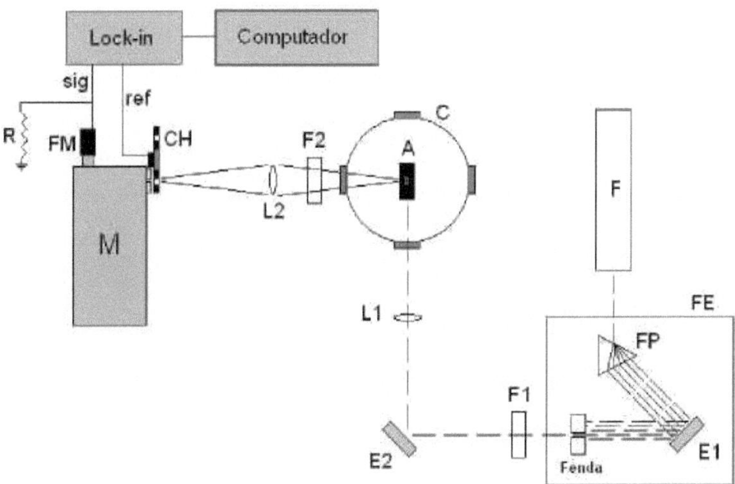

Figure 8 Representative diagram of the system used for luminescence measurements.

The beam passes through filter *F1* (a low-pass filter that allows light with wavelengths shorter than 500.0 nm to pass through). Mirror *E2* directs the beam towards converging lens *L1*, which focuses the light on sample *A*. The luminescence is collected by lens *L2,* which focuses it on the entrance slit of monochromator *M.* Filter F2 is now a 510 nm high-pass filter that allows the luminescence light to pass through. From now on, the same situation is repeated as in the case of the *upconversion* measurement system.

Chapter 4

4.   *Results and discussion*

4.1,   *Upconversion results in the* $^4S_{3/2}$ → $^4I_{15/2}$ *transition for the* ZBLAN:Er^{3+} *system at T=2K*

4.1.1.   *Result at T=2K*

Table 3 in section 2.2.1 shows that the multiplets $^4S_{3/2}$ e $^4I_{15/2}$ have two and eight

components respectively. At T=2K the transitions start from the lower energy level of the

multiplet S$^4_{3/2}$ to the eight energy levels of the fundamental multiplet, which are the eight

expected transitions, and which are shown in figure 9.

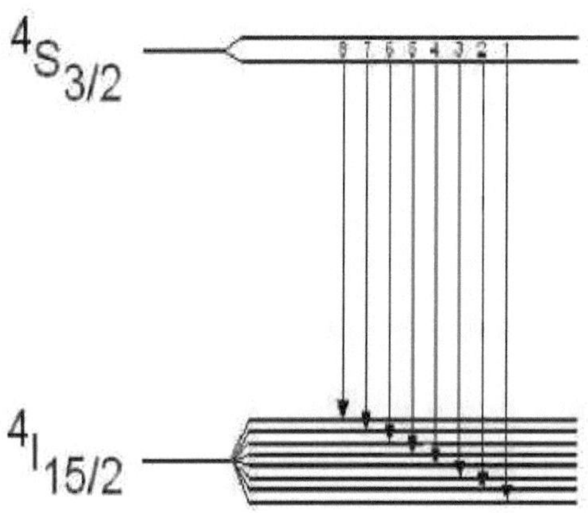

Figure 9 Schematic representation of the expected transitions.

Figure 10 shows the *upconversion* emission spectrum for the ZBLAN:Er^{3+} matrix

at a temperature of 2K. The excitation was carried out using a Titanium Sapphire laser whose

emitted light was tuned to a wavelength of 800 nm (energy 12500 cm^{-1} , corresponding to the

$^4I_{15/2}$ → $^4I_{9/2}$). transition. An analysis of the figure shows that
Only seven emission lines are observed, out of the eight theoretically expected.

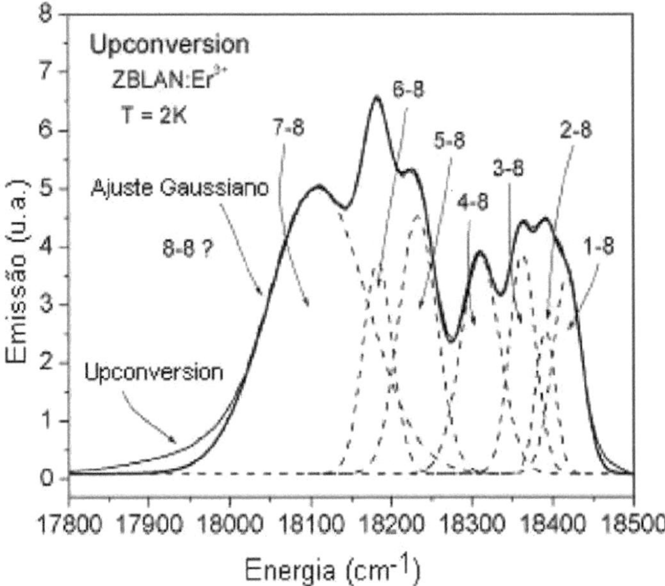

Figure 10. *Upconversion* spectrum obtained at T=2K in the ZBLAN system: Er^{3+} for the transition $^4S_{3/2} \rightarrow {}^4I_{15/2}$. Resolution 0.08 Å and Error < 0.01%. Line positions (cm^{-1}): 18416, 18387, 18349, 18302, 18236, 18186 and 18117.

For the *upconversion* process to be possible, electronic transitions between the energy levels of an ion must take place in a certain sequence. In the case under study, the excitation of the *upconversion* emission can be induced in 3 different ways, which will be presented below.

The sequence in figure 11 was considered as possibility 1. The system sequentially absorbs 2 photons of energy equal to 12500 cm^{-1} . In the figure we see that the first photon is absorbed in a state transition $^4I_{15/2} \rightarrow {}^4I_{9/2}$ (1). A second photon is absorbed, causing the ion to reach the multiplet $H^2{}_{9/2}$ (2).

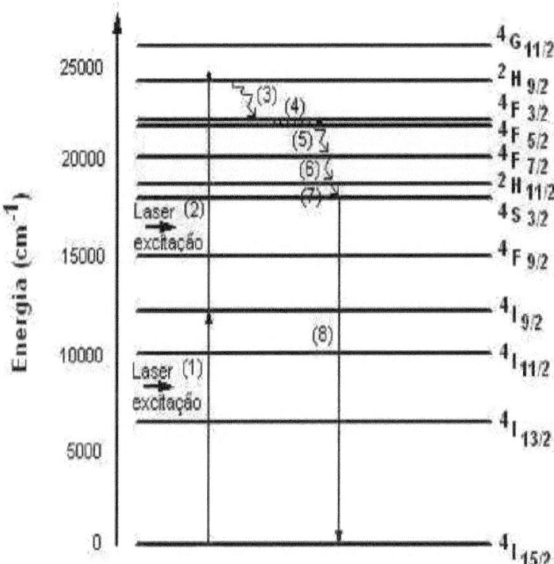

Figure 11. Alternative diagram representing the energy levels for the Er ion *upconversion* process[3+].

The ion would then reach the level S $/^4{}_{32}$ through successive non-radiative decays (3), (4), (5), (6) and (7). From there it would transition to the ground state (8), emitting the spectrum shown in figure 10.

Another sequence, identified as possibility 2, is represented schematically in figure 12. In this case, the ion would transition from the level $I^4{}_{15/2}$ to the level $I^4{}_{9/2}$ (1) by absorbing an excitation photon. It would then decay non-radiatively to the level $I^4{}_{11/2}$ (2). By absorbing a second photon from the exciter beam, it would transition to the level F $/^4{}_{32}$ (3). Now, by non-radiative processes, it would decay sequentially to the level $S^4{}_{3/2}$ (4), (5), (6), and (7). From there, it would transition to the ground state (8), emitting the spectrum shown in figure 10.

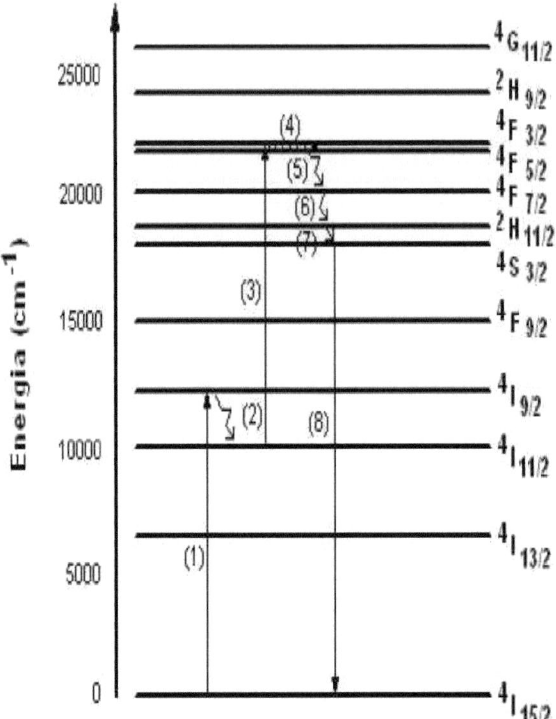

Figure 12. Alternative diagram representing the energy levels for the Er ion *upeonversion* process[3+] .

Lastly, possibility 3, shown in figure 13, will be described. The ion would transition from the level I $/^4{}_{152}$ to the level I $/^4{}_{92}$ (1) by absorbing an excitation photon. Then, through non-radiative decay, it would move to the level $I^4{}_{11/2}$ (2) and then to the level $I^4{}_{13/2}$ (3). By absorbing another photon from the excitation source, it would pass to the level $H^2{}_{11/2}$ (4) from where it would decay non-radiatively to the level $S^4{}_{3/2}$ (5). From there it would transition to the ground state (6), emitting the spectrum shown in figure 10.

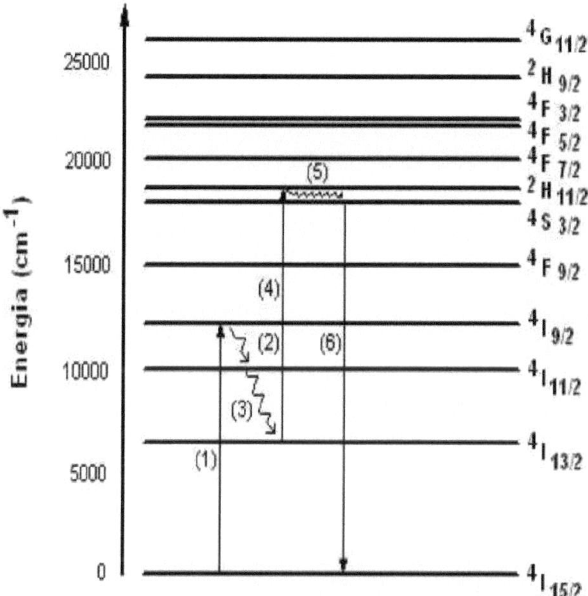

Figure 13. Alternative diagram representing the energy levels for the Er ion *upconversion* process[3+] .

Possibilities 2 and 3 can be considered as probable, and are justified by the experimental non-appearance of the emissions for the higher energy transitions, i.e.: $^{4}F_{5/2} \rightarrow\, ^{4}I_{15/2}$, $^{4}F_{7/2} \rightarrow\, ^{4}I_{15/2}$. To reinforce this suspicion also

no emission was observed in the transition $^{2}H_{11/2} \rightarrow\, ^{4}I_{15/2}$. To help define the *upconversion* process, a search was made for references in international literature.

N. Jaba et al [4] published a paper on upconversion of energy at wavelengths in the infrared to visible range in Er-doped glasses[3+] . Figure 14 shows the energy level diagram representing the possible transitions for the *upconversion* process presented by the authors. They define the following for the figure: dotted line representing the *upconversion* pump, continuous line for radiative emission and curved line for non-radiative emission.

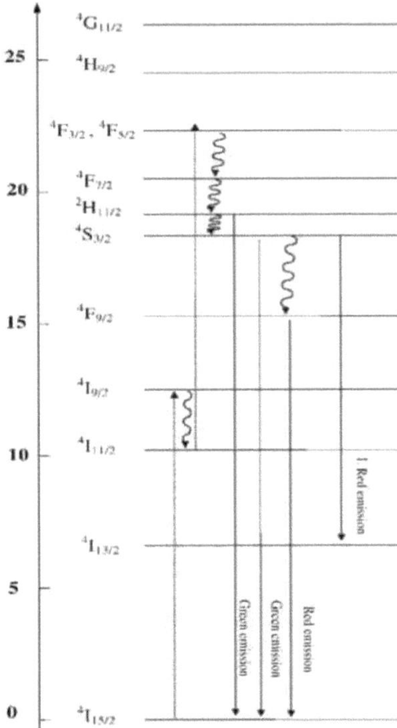

Figure 14. Diagram of energy levels representing the possible transitions for the *upconversion* process in glass doped with Er^{3+} [4].

Analysis of the figure shows that an absorption transition takes place, with the ion moving from the level $I^4_{15/2}$ to the level $I/^4{}_{92}$ by absorbing an excited photon. It then decays non-radiatively to the level $Ih^4{}_{2}$. By absorbing a second photon from the exciter beam, it transitions to the level $F^4{}_{3/2}$ and/or $F^4{}_{5/2}$. Now, by non-radiative processes, it decays sequentially to the level $S^4{}_{3/2}$. From there, it transitions to the ground state with the emission of photons in the green region of the electromagnetic spectrum. This process defined by N. Jaba et al is similar to possibility 2 (figure 12) described in section 4.1.1. Figure 15 shows the luminescence spectrum obtained by the authors.

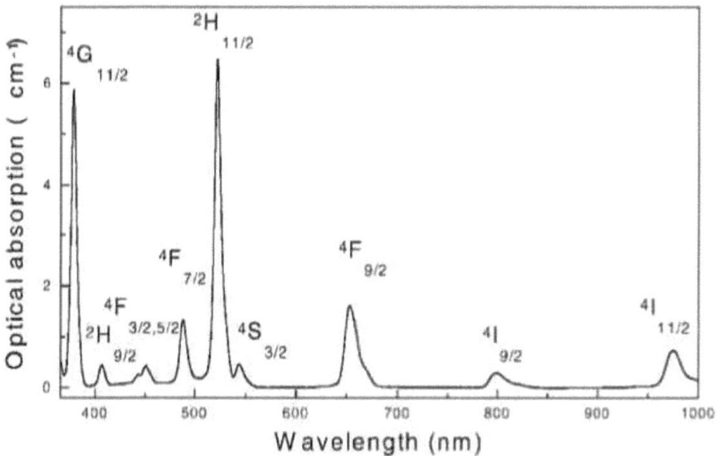

Figure 15. Luminescence spectrum obtained at T=300K for a 70TeO$_2$ -30ZnO glass matrix with 0.4% Er^3 ' showing the transition line ^{4}S$_{3/2}$ → ^{4}I$_{15/2}$ and excitation at wavelength 797 nm [4].

Another work worth mentioning is that published by M. Tsuda et al [11], whose topic is directly related to that of this dissertation, as it deals with the study of the mechanism of the *upconversion* phenomenon in fluorozirconate glasses doped with Er^{3+} at 800 nm excitation. Figure 16 transcribes one of the possible transitions for *upconversion* described by the authors. In this figure it can be seen that photon absorption from the exciter beam occurs with the ion transitioning from the level I^4$_{i5/2}$ to the level I^4$_{9/2}$ (GSA). Then, through the absorption of a second photon from the exciter beam, it transitions to the level H^2$_{9/2}$ (ESA). Now, probably by non-radiative processes, it decays sequentially to the level S^4$_{3/2}$. From there it transitions to the ground state with the emission of photons from the green visible region (550 nm) of the electromagnetic spectrum.

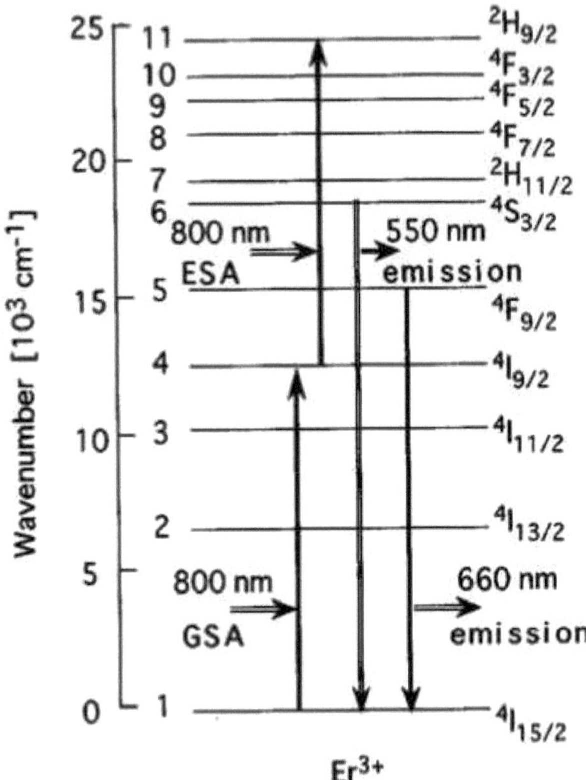

Figure 16 Schematic diagram of energy levels for the Er ion³ ' [11].

This process defined by M. Tsuda et al is similar to possibility 1 (figure 11)   Figure 17 shows the luminescence spectrum obtained by the authors.

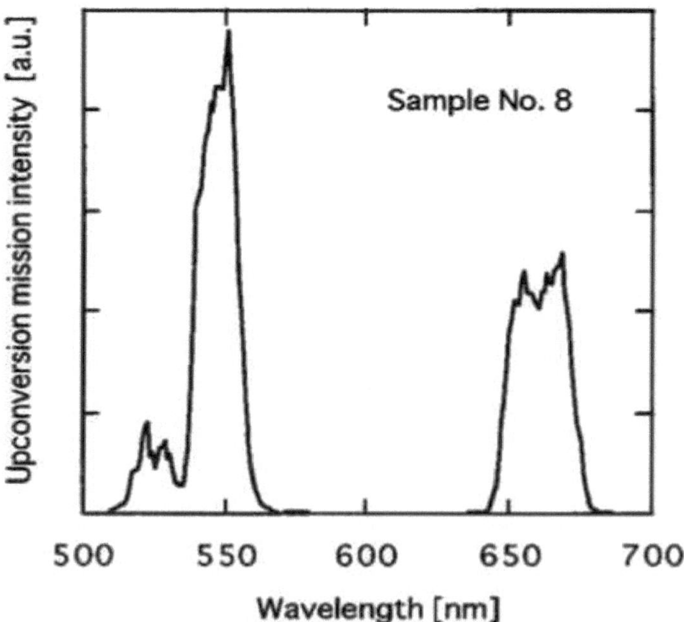

Figure 17 Luminescence spectrum obtained at room temperature for a ZBLAN glass matrix (Sample 8
→ 47.4% ZrF_4 – 17.9% BaF_2 - 3.6% LaF_3 - 2.7% AlF_3 - 17.9% NaF - 0.5% InF_3 - 10% ErF_3) showing the
transition line $^4S_{3/2}$ → $^4I_{15/2}$ at 550 nm. Excitation at a wavelength of 800 nm [11].

M. In the same paper, Tsuda et al also describe what would be the dominant
mechanism of the *upconversion* process in ZBLAN:Er^{3+} , which is shown in figure 18. An
analysis of the figure and the description of this mechanism cited by the authors shows that
the process for *upconversion* emission at 550 nm with excitation at 800nm occurs as follows:
Er^{3+} is excited from level 1 to 4 (GSA) which is equivalent to the transition between the levels
$^4I_{15/2}$ → $^4I_{9/2}$, after which it decays to the state
$^4I_{11/2}$ by a non-radiative process (MPR). The transition $^4I_{11/2}$ → $^4F_{7/2}$ is excited by
energy transfer from a transition $^4I_{11/2}$ → $^4I_{15/2}$. Decay of the
level $^4F_{7/2}$ até o $^4S_{3/2}$ by successive non-radiative processes (MPR) and then occurs the
transition $^4S_{3/2}$ → $^4I_{15/2}$ with 550 nm emission. Note that
some transitions that occur by energy transfer. This process defined by M. Tsuda et al. is
similar to possibility 3 (figure 13) described in section 4.1.1. as one of the possible transitions

for *upconversion to* occur.

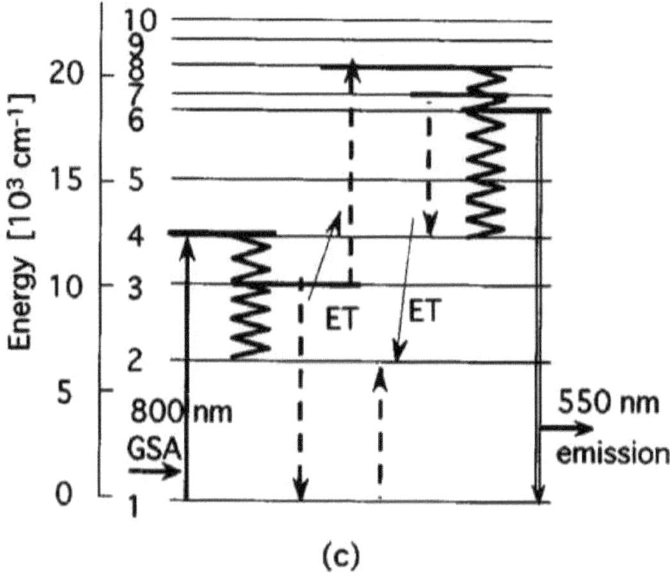

(c)

Figure 18. Schematic diagram of the energy levels and dominant mechanism for the *upconversion* process in Er-doped fluorozircon glass[3+] with excitation at 800 nm [11].

Another work that attracts attention is that published by S. Q. Man et al [10] which also deals with the study of *upconversion* luminescence in glass doped with Er^{3+} . Figure 19 reproduces the diagram that the authors determined as the transitions that occur for the *upconversion* phenomenon in their sample.

In this figure it can be seen that the ion transitions from the level I $/^4{}_{152}$ to the level I $/^4{}_{92}$ by absorbing the excitation photon. It then decays non-radiatively to the level $I^4{}_{11/2}$. By absorbing a second photon from the exciter beam, it transitions to the level $F^4{}_{3/2}$. Now, by non-radiative processes, it decays sequentially to the level $S^4{}_{3/2}$. From there, it transitions to the ground state with green emission at 547 nm.

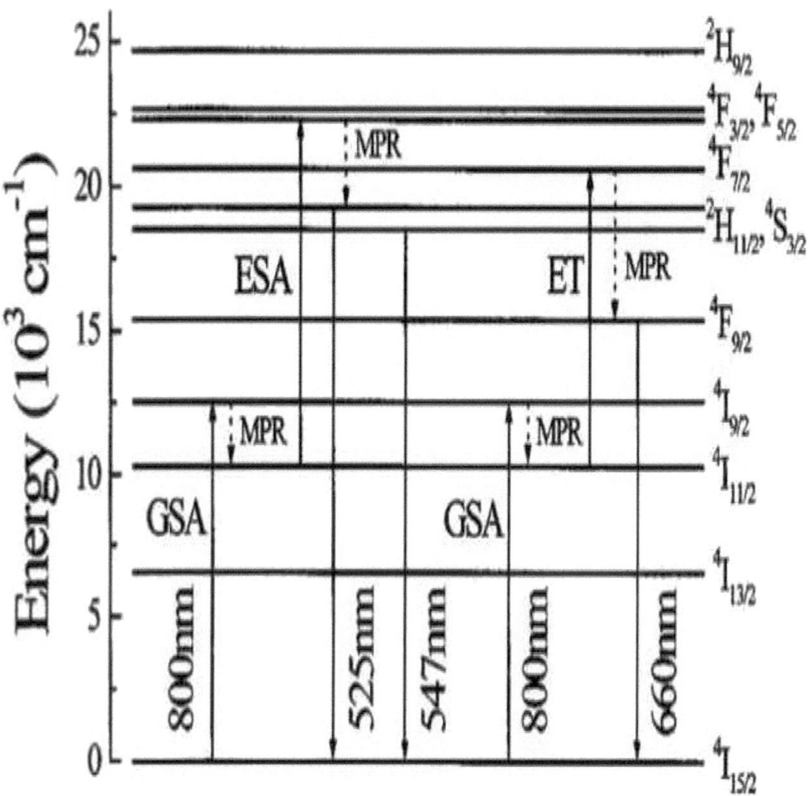

Figure 19. Energy level diagram representing the transition mechanism for the *upconversion* process in KGB glass matrix ($5K_2O$ - $70Bi\,O_{23}$ - $25Ga\,O_{23}$) doped with Er^{3+} for emissions at 525, 547 and 660 nm [10].

This work by S. Q. Man et al is similar to possibility 2 in figure 12 of section 4.1.1.

Figure 20 shows the luminescence spectrum obtained by the authors.

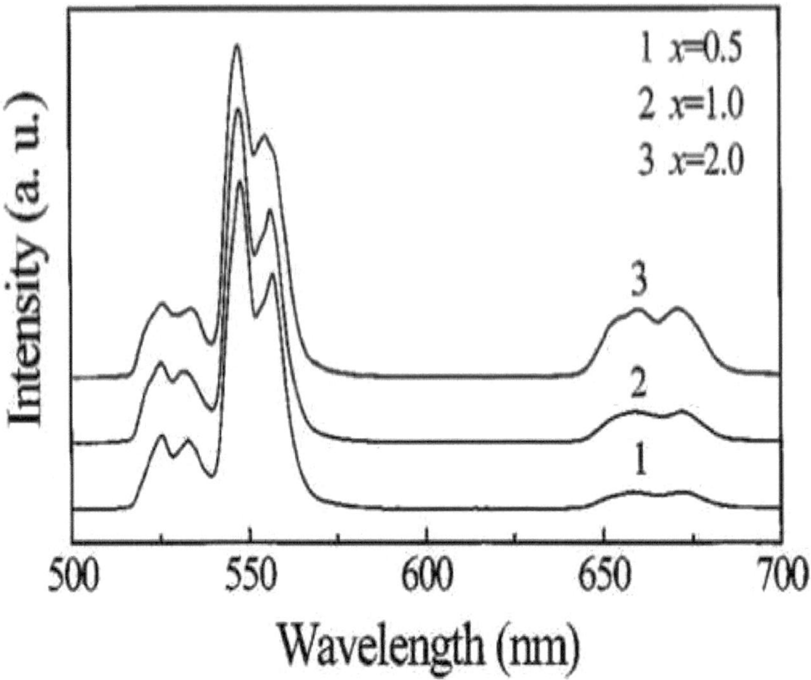

Figure 20. Luminescence spectra for KGB glass matrix → $5K_2O$–$70Bi_2O_3$–$25Ga_2O_3$:xEr_2O_3 (x=0.5, 1, 2). Emission at 550 nm is evident [10].

Based on the research carried out to help define the *upconversion* process, it can be seen that of the 4 results presented and described above, in 3 of them we have confirmation of the transitions in possibilities 2 and 3 of figures 12 and 13 proposed in item 4.1.1.

It can be seen from the spectra presented that none of them were obtained at low temperatures. Therefore, experimental results were sought under the conditions of interest in this work, i.e. low temperatures. In this respect, it was found that at temperatures below ambient T there are no records for this system for the *upconversion* process, but for comparison, we used a luminescence spectrum obtained at T=2K which we will now present since the emission transition investigated for both cases is the same, i.e., $^4S_{3/2} \rightarrow {}^4I_{15/2}$.

4.2. *Luminescence results*

4.2.1. *Emission results in the $S^4{}_{3/2}$*
$\rightarrow {}^4I_{15/2}$ *transition for the ZBLAN:Er³⁺ system at T=2K*

Figure 21 shows the luminescence spectrum for the ZBLAN:Er³⁺ matrix at a

temperature of 2K [17]. The excitation was carried out with an Argonium laser emitting light

with a wavelength of 488 nm (energy of 20491 cm⁻¹ , corresponding to the transition

$^4S_{3/2} \rightarrow {}^4F_{7/2}$). In the spectrum of the figure, eight
lines, resulting from the eight transitions between the lowest energy level of the multiplet S

$/^4{}_{32}$ and the levels of the fundamental multiplet $I^4{}_{i5/2}$, a theoretically expected result.

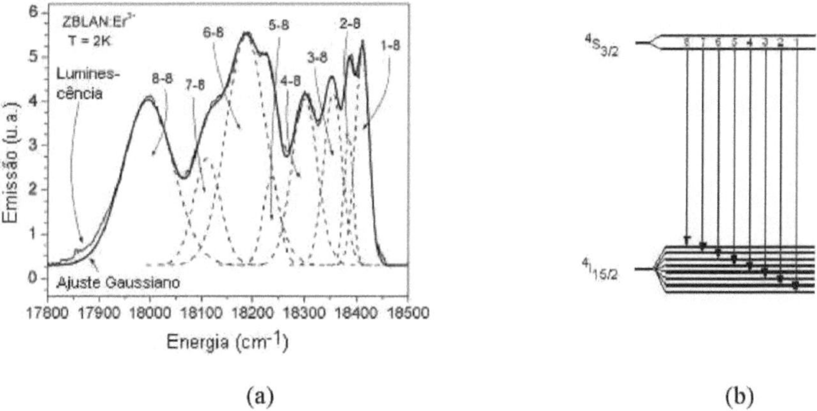

(a)   (b)

Figure 21. (a) Luminescence spectrum at T=2K for the ZBLAN:Er³⁺ system shown in solid line. The eight peaks expected with the help of Gaussian decomposition appear in dashed line [18]. (b) Representation of the expected transitions $^4S_{3/2} \rightarrow {}^4I_{15/2}$. Resolution 0.08 Á and Error < 0.01%. Position of
lines (cm⁻¹): 18416, 18387, 18349, 18302, 18236, 18186, 18117 and 17996.

    This spectrum in the figure, in which the eight lines of transitions that allow the

positions of the eight energy levels of the multiplet⁴ ₁ₛ/₂ of the Er ion³⁺ in the ZBLAN matrix

can be easily identified, should be equivalent to that realized in the *upconversion* emission.

However, this is not the case.

By comparing the *upconversion* and luminescence emission spectra in figure 22, it can be seen that
the lowest energy line is not present in the *upconversion* spectrum (line 8).

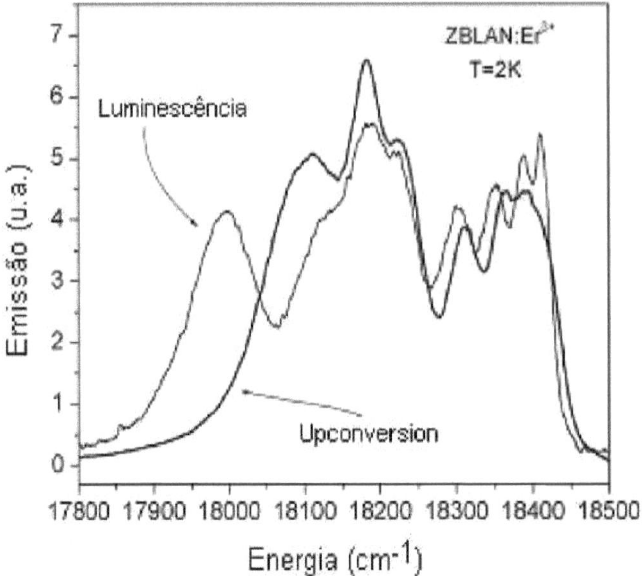

Figure 22. Comparison between *upconversion* emission and luminescence emission at 2K.

Next, the evolution of the luminescence emission in the $^4S_{3/2} \rightarrow \,^4I_{15/2}$ transition was monitored with temperature to verify the behavior of the emission spectrum obtained with the change in temperature. The results are presented below.

4.2.2 Emission results in the transition $S^4_{3/2} \rightarrow \,^4I_{15/2}$ for the ZBLAN:Er system^{3+} at T>2K - comparison

Figure 23 shows 3 luminescence spectra in the ZBLAN: Er^{3+} system for 3 different temperatures [17]. Nothing can be said about the temperature of spectra (2) and (3) since the cooling system did not allow this information to be obtained.

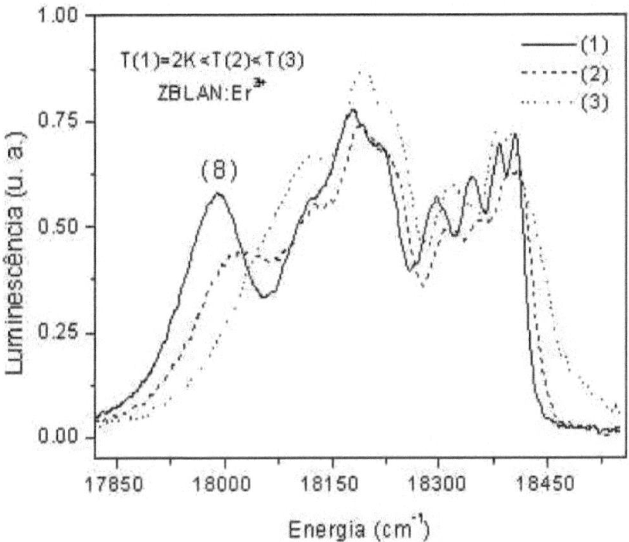

Figure 23. Luminescence spectrum for the $^{4}S_{3/2} \rightarrow {}^{4}I_{15/2}$ transition in the ZBLAN:Er^{3+} system obtained for different temperatures. T1=2K. T2 and T3 are intermediate temperatures between 2 and 15 K. Resolution 0.08 Á and Error < 0.01%.

However, spectrum (1) was obtained at T = 2K and spectra (2) and (3) were obtained by waiting for the cooling system to spontaneously raise its temperature.

It can be seen that the luminescence spectrum identified with the number (3) is similar to the emission spectrum obtained by the *upconversion* process at T = 2K (figure 10). Therefore, if the final state of the transition is the same, it can be inferred that in the *upconversion* process, part of the energy is somehow absorbed by the system. It should be remembered that the difference between the luminescence and *upconversion* processes is only the excitation.

Figure 22 compares the *upconversion* emission with the luminescence emission at a temperature of 2K. The comparison shows that the *upconversion* emission at T=2K has a spectrum profile similar to that of the luminescence emission at a higher temperature (figure 23 - curve 3). In the light of these results, we searched for experimental low-temperature luminescence results in the literature, which will be presented in the next section.

4.3. Low-temperature luminescence results obtained from the literature

In an article published in 2001 [18], Y. D. Huang et. al. presented the luminescence spectrum for the $^4S_{3/2} \rightarrow {}^4I_{15/2}$ transition of the Er ion^{3+} in the ZBLAN matrix. at T=13K. The experimental result of the article is transcribed in figure 24.

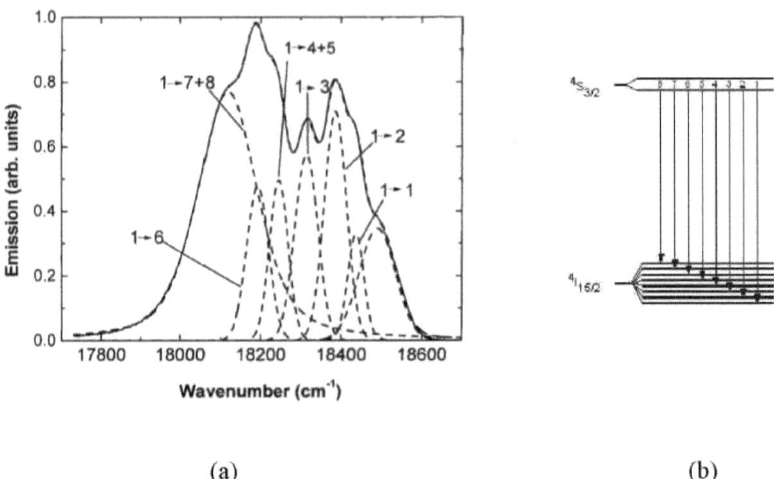

(a)   (b)

Figure 24. (a) Luminescence spectrum of the $^4S_{3/2} \rightarrow {}^4I_{15/2}$ transition for the ZBLAN:Er^{3+} matrix at T=13K [18].
(b) Representation of the expected transitions I $/^4S_{3/2} \rightarrow {}^4_{152}$. Note: Gaussian not shown (before number 1)
this is the duplication of lines due to the increase in temperature.

The authors identified the eight lines that would be expected for this transition.

Table 4 represents part of the results in which Huang et. al. present the indexing of the positions of the eight and two energy levels of the multiplets$^{4I}{}_{15/2}$e ^{4S} 3/2 respectively.

Table 4: Transcription of the energy positions (cm^{-1}) in accumulated value for the Stark levels of the multiplets^4 s3/2e4l 15/2-

Manifold	Experimental
	Level position$^{'1}$
	ü
	47

	111
$^4I_{15/2}$	154
	194
	226
	308(√)
	3O8(^)
$^4S_{3/2}$	18.429
	18.525

Source: Bibliographic reference [18].

The article cited was the only one found in the bibliography that presents luminescence results for ZBLAN: Er^{3+} , in the transition in $^4S_{3/2} \rightarrow {}^4I_{15/2,}$ temperatures below 15K. However, it can be seen that there is a discrepancy between the spectrum and the results of the data table presented. Figure 24 shows a spectrum adjusted using the six Gaussians indicated, but table 4 shows eight positions, two of which have been given the same value. Analyzing the aperture of the multiplet I $/^4{}_{152}$ shows that table 4 indicates an aperture of 308 cm^{-1} , while the experimental result leads to an aperture of 338 cm^{-1} for the same multiplet, verified graphically.

Despite this discrepancy between values, it can be seen that the spectrum obtained by Huang is similar to the one in figure 10. This comparison is possible since it is the same emission transition. Our suspicion is that the spectrum (3) in figure 22 was obtained close to the temperature of the spectrum obtained by Huang et al. This assumption is made considering that the profiles of the spectra are similar.

4.4. Final discussions

Emission spectra from *upconversion* and luminescence at 2K were compared (figure 22). From this comparison, it was observed that in the *upconversion* emission process at a temperature of 2K, there must be a process that leads to the absence of the expected eighth line (which has an energy of around 18000cm^{-1}) in the transition. $^4S_{3/2} \rightarrow {}^4I_{15/2}$.

Tables 5a and 5b help to justify our assumption. They were compiled from the free ion energy values listed in table 2 and indicate all possible values of the differences between the Er ion energy levels[3+] .

Table 5a. Differences in values between the energy levels for the Er ion[3+] .

Indice	Nível	Energia (cm^{-1})	ΔE (rel $^4I_{15/2}$)	ΔE (rel $^4I_{13/2}$)	ΔE (rel $^4I_{11/2}$)	ΔE (rel $^4I_{9/2}$)	ΔE (rel $^4F_{9/2}$)	ΔE (rel $^4S_{3/2}$)
1	$^4I_{15/2}$	0	0	-6485	-10123	-12345	-15182	-18299
2	$^4I_{13/2}$	6485	6485	0	-3638	-5860	-8697	-11814
3	$^4I_{11/2}$	10123	10123	3638	0	-2222	-5059	-8176
4	$^4I_{9/2}$	12345	12345	5860	2222	0	-2837	-5954
5	$^4F_{9/2}$	15182	15182	8697	5059	2837	0	-3117
6	$^4S_{3/2}$	18299	18299	11814	8176	5954	3117	0
7	$^2H_{11/2}$	19010	19010	12525	8887	6665	3828	711
8	$^4F_{7/2}$	20494	20494	14009	10371	8149	5312	2195
9	$^4F_{5/2}$	22181	22181	15696	12058	9836	6999	3882
10	$^4F_{3/2}$	22453	22453	15968	12330	10108	7271	4154
11	$^2H_{9/2}$	24475	24475	**17990**	14352	12130	9293	6176
12	$^4G_{11/2}$	26376	26376	19891	16253	14031	11194	8077

Source: Data calculated based on Table 2.

Table 5b. Differences in values between energy levels for the Er ion[3+] .

Indice	Nivel	Energia (cm^{-1})	ΔE (rel $^2H_{11/2}$)	ΔE (rel $^4F_{7/2}$)	ΔE (rel $^4F_{5/2}$)	ΔE (rel $^4F_{3/2}$)	ΔE (rel $^2H_{9/2}$)	ΔE (rel $^4G_{11/2}$)
1	$^4I_{15/2}$	0	-19010	-20494	-22181	-22453	-24475	-26376
2	$^4I_{13/2}$	6485	-12525	-14009	-15696	-15968	-17990	-19891
3	$^4I_{11/2}$	10123	-8887	-10371	-12058	-12330	-14352	-16253
4	$^4I_{9/2}$	12345	-6665	-8149	-9836	-10108	-12130	-14031
5	$^4F_{9/2}$	15182	-3828	-5312	-6999	-7271	-9293	-11194
6	$^4S_{3/2}$	18299	-711	-2195	-3882	-4154	-6176	-8077
7	$^2H_{11/2}$	19010	0	-1484	-3171	-3443	-5465	-7366
8	$^4F_{7/2}$	20494	1484	0	-1687	-1959	-3981	-5882
9	$^4F_{5/2}$	22181	3171	1687	0	-272	-2294	-4195
10	$^4F_{3/2}$	22453	3443	1959	272	0	-2022	-3923
11	$^2H_{9/2}$	24475	5465	3981	2294	2022	0	-1901
12	$^4G_{11/2}$	26376	7366	5882	4195	3923	1901	0

Source: Data calculated based on Table 2.

Table 5a shows that the $^4I_{13/2} \rightarrow {}^2H_{9/2}$ transfer involves an energy absorption of around 17990 cm^{-1} , which directly justifies our suggested energy reabsorption process.

Therefore, based on this experimental observation, the *upconversion* excitation process shown schematically in figure 25 is suggested.

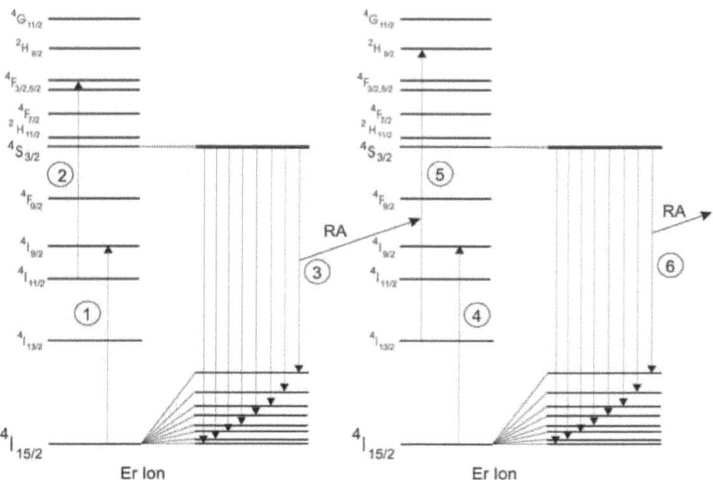

Figure 25. Representative diagram of the energy reabsorption of line 8 of the $^4S_{3/2} \rightarrow {}^4I_{15/2}$ transition in the system ZBLAN:Er^{3+} .

The first part of the process is as shown in figure 12, i.e. the absorption of a photon from the excitation beam which causes the system to transition from the state $I^4{}_{15/2}$ to the state $I^4{}_{9/2}$. The ion then decays non-radiatively to the state $I^4{}_{11/2}$ from which it transitions to the state F ,$^4{}_{3/25/2}$ through a new absorption of a photon from the excitation beam. Through non-radiative decays, it reaches the level $S^4{}_{3/2}$ from where the emission transitions of the luminescence spectrum shown in figure 21 will start. In the case of the *upconversion* process, what can happen is that the lower energy transition can be "reabsorbed" by a neighboring ion which is in the $I^4{}_{13/2}$ state after having absorbed an exciton photon ($I^4{}_{15/2}$ $\rightarrow$ $_4$ $I_{9/2}$) and having decayed non-radiatively between the I $/2^4{}_9{}^e$ ^{4I} 13/2- levels. This reabsorption causes the neighboring ion to transition to the state $H^2{}_{9/2}$ from which it returns to the state $S^4{}_{3/2}$ by non-radiative processes. A new emission can occur and the process is repeated in this internal energy feedback. As a consequence of this process of energy transfer between neighboring ions through reabsorption, the *upconversion* emission spectrum shown in figure 10 is the result.

Tables 5a and 5b have been compiled with all the possible difference values for the energy levels of Er^{3+} in the possible transitions that could occur. Having checked it, we can see that the probable energy value for the absorption of the energy value of approximately 18,000 cm^{-1} would be the transition $^4I_{13/2}$ $\rightarrow$ $^2H_{9/2}$.

Conclusions

Figure 10 shows the *upconversion* emission spectrum at 2K. There are seven of the eight lines expected for transinoes between the lowest energy level of the multiplet $S/^4_{32}$ and the fundamental multiplet.

In the literature, there were no emission results from the low-temperature *upconversion* process that were similar to those presented in this work. So we compared the *upconversion* result with the luminescence result at T=2K, the latter showing the eight lines expected for transinoes between the lowest energy level of the multiplet $S^4_{3/2}$ and the fundamental multiplet.

In order to confirm the results, which would lead to confirmation of the measurement process used, we searched the literature for low-temperature luminescence results. From this search, we obtained the work published by Huang et. al with a spectrum obtained at T=13K. This comparison confirmed that the measurements carried out in this work were correctly conducted.

Finally, 3 possible excitation paths were proposed for the *upconversion* process and from there we searched the literature for confirmation of the occurrence of these paths. If 2 of them were confirmed, along with the absence of the expected eighth line (which has an energy of around 18000cm^{-1}) in the transition $S/^4_{32} \rightarrow {}^4I_{15/2}$, we arrived at the composition and defined that the levels involved and the *upconversion* emission process should be the one shown and discussed in figure 25 with the occurrence of the energy reabsorption process.

Bibliography

[1]   W. Lozano B. et al, Upconversion of infrared-to-visible light in Pr -Yb^{3+3+} codoped fluorindate glass, Optics Communications v. 153, p. 271-274 (1998).

[2]   K. Soga et al, Effects of chloride introduction on the optical properties and the upconversion emission with 980-nm excitation of Er^{3+} in ZBLAN fluoride glasses, Journal of Non-Crystalline Solids v. 222, p. 272-281(1997).

[3]   Kohei Soga et al., Calculation and simulation of spectroscopic properties for rare earth ions in chloro-fluorzirconate glasses, Journal of Non-Crystalline Solids v. 274, p. 69-74 (2000).

[4]   N. Jaba et al, Infrared to visible up-conversion study for erbium-doped zinc tellurite glasses, Journal of Physics: Condensed Matter v. 12, p. 4523-4534 (2000).

[5]   M. Tsuda et al, Effect of Yb^{3+} doping on upconversion emission intensity and mechanism in Er /Yb^{3+3+} -codoped fluorzirconate glasses under 800nm excitation, Journal of Applied Physics v. 86 n. 11, p. 6143-6149 (1999).

[6]   Chen Xiaobo, Study of up-conversion luminescence of Er(2):ZBLAN excited by 1520 nm laser, Optics Communications v. 242, p. 565-573 (2004).

[7]   Hongtao Sun et al., Letter to the Editor: Efficient frequency upconversion emission in Er^{3+} -doped novel strontium lead bismuth glass, Journal of Non-Crystalline Solids v. 351, p. 288-292 (2005).

[8]   Z. Meng et al, Energy transfer mechanism in Yb^{3+} :Er^{3+} -ZBLAN: macro- and micro-parameters, Journal of Luminescence v. 106, p. 187-194 (2004).

[9]   Yanbing Hou et al, Temperature dependence of upconversion luminescence from Pr^{3+} and Yb^{3+} codoped ZBLAN glass pumped by 960 nm laser diode, Journal of Non-Crystalline Solids v. 260, p. 54-58 (1999).

[10] S. Q. Man et al, Upconversion luminescence of Er^{3+} in alkali bismuth glasses,

Journal of Applied Physics v. 77 n. 4, p. 483-485 (2000).

[11] M. Tsuda et al, Upconversion mechanism in Er^{3+} -doped fluorzirconate glasses under 800 nm excitation, Journal of Applied Physics v. 85 n. 1, p. 29-37 (1999).

[12] A.S. Oliveira et al, Upconversion fluorescence spectroscopy of Er /Yb^{3+3+} -doped heavy metal Bi2O3-Na2O-Nb2O5-GeO2 glass, Journal of Applied Physics v. 83 n. 1, p. 604-606 (1998).

[13] L. D. da Vila et al, Mechanism of the Yb-Er energy transfer in fluorzirconate glass, Journal of Applied Physics v. 93 n. 7, p. 3873-3880 (2003).

[14] Ahmed Boutarfaia et al, New stable fluoride glasses, Solid State Ionics v. 144, p. 117-121 (2001).

[15] Y. C. Yan et al, Erbium-doped phosphate glass waveguide on silicon with 4.1 dB/cm gain at 1.535 pm, Applied Physic Letters v. 71 n. 20, p. 2922-2924 (1997).

[16] J. S. Sanghera, Infrared Optical Fibers, edited by J. S. Sanghera and I. Aggarwal, CRC Press, New York (1998).

[17] Viatroski, M. A., Spectroscopic analysis of the vitreous system ZBLAN:Er^{3+} at low temperature, Dissertation (Master's Degree in Materials Science and Engineering), Interunits Materials Science and Engineering, USP (2003).

[18] Y. D. Huang et al, Stark level analysis for Er^{3+} -doped ZBLAN glass, Optical Materials v. 17, p. 501-511 (2001).

[19] V. D. Rodríguez et al, Spectroscopy of rare earth ions in fluoride glasses for laser applications, Optical Materials v. 13, p. 1-7 (1999).

[20] M. Mortier et al, Crystal field analysis of Er^{3+} -doped glasses: germanate, silicate and ZBLAN, Journal of Alloys and Compounds v. 300-301, p. 407-413 (2000).

[21] A. Florez et al, Optical transition probabilities and compositional dependence of

Judd-Ofelt parameters of Nd^{3+} ions in fluorindate glasses, Journal of NonCrystalline Solids v. 284, p. 261-267 (2001).

[22] X.B.Chen et al, The comparison investigation of direct upconversion sensitization luminescence between ErYb:oxyfluoride glass and vitroceramics, Optics Communications v. 184, p. 289-304 (2000).

[23] S. Hüfner, Optical Spectra of Transparent Rare Earth Compounds, Academic Press, New York (1978).

[24] G. K. Cruz, Optical properties of the Er ion Y^{3+}_2 $BaZnO_5$, Thesis (Doctorate in Sciences), IFSC-USP (1998).

[25] A. B. Londoño, Estudo das propriedades ópticas dos vidros fluoroindatos dopados com Er^{3+} e Pr^{3+} , Tese (Doutorado em Ciências e Engenharia de Materiais), Interunidades Ciência e Engenharia de Materiais, USP (1996).

[26] B. S. Cardiel, Caracterización de defectos intrínsicos y extrínsicos en zirconia estabilizada con ytria, Tesis (Doctorado en Ciências Físicas), Departamento de Física de la Universidad Carlos III de Madrid (1998).

[27]   Bartolo, B. D., Optical Interactions in Solids, John Wiley & Sons, Inc. (1968).

Printed by Books on Demand GmbH, Norderstedt / Germany